Frederick Edward Hulme

The Garland of the year

The months - their poetry and flowers with twelve chromographs of

flowers, one for each month

Frederick Edward Hulme

The Garland of the year
The months - their poetry and flowers with twelve chromographs of flowers, one for each month

ISBN/EAN: 9783337815967

Printed in Europe, USA, Canada, Australia, Japan

Cover: Foto ©berggeist007 / pixelio.de

More available books at **www.hansebooks.com**

The Garland of the Year.

THE

GARLAND OF THE YEAR;

OR,

The Months: their Poetry and Flowers.

WITH

TWELVE CHROMOGRAPHS OF FLOWERS,

ONE FOR EACH MONTH.

" Blessed be God for flowers ;
For the bright, gentle, holy thoughts, that breathe
From out their odorous beauty, like a wreath
Of sunshine on Life's hours."

MRS. TINSLEY.

London :

MARCUS WARD & CO., CHANDOS ST., W.C. ;
AND ROYAL ULSTER WORKS, BELFAST.

PRINTED BY

MARCUS WARD AND CO.

ROYAL ULSTER WORKS

BELFAST

PREFACE.

THE Editor of this little work has endeavoured to make his descriptions of the months, which are necessarily very brief, both interesting and instructive. For much of the historical information embodied in these, he is indebted to the late Dr. Robert Chambers's admirable "Book of the Year." The colored illustrations are by Mr. F. Edward Hulme, F.L.S., F.S.A.

The chief poetical selections are carefully chosen from our Standard British Poets. Some are original: all relate specially to the months and their flowers. Pains have been taken to render the work suitable for a gift book, at any season, or for a birthday present, on any day of the year.

Sᴇᴇ the minutes how they run ;
How many make the hour full complete,
How many hours bring about the day,
How many days will finish up the *year*,
How many *years* a mortal man may live !

Shakespeare.—Henry VI.

The Garland of the Year.

CONTENTS.

4　　　　　　　　　*Contents.*

SNOWDROP

JANUARY

Emblem of purity,
       white as the snow,
Among which so often
       your blossoms grow,
Bringing back life
       amid Winter's decay,
Braving the blast
       of a cold frosty day ;
Decking with beauty
       the garden then drear
First of the flowers
       of opening year.

THE YEAR OF MAN'S LIFE.

"Though we seem grieved at the shortness of life in general, we are wishing every period of it at an end. The minor longs to be at age ; then to be a man of business ; then to make up an estate ; then to arrive at honors ; then to retire."—ADDISON.

THE birth, life, and death of the year, and of the flowers, have often been used to illustrate the shortness of life, and the certainty of death ; for all must fade and die. The idea has been fitly described in the annexed quaint old poem, written by an unknown author, in 1653.

JANUARY.

THE first five years then of man's life,
 Compare to Januar ;
In all that time but sturt and strife,
 He can but greet and roar ;
So in the fields of flowers all bare,
 By reason of the frost :
Keeping the ground both soft and sound,
 Yet none of them is lost.

FEBRUARY.

So to ten years I shall speak then,
 Of Februar but lack ;
The child is meek and weak of spirit,
 Nothing can undertake.

So all the flowers, for lack of showers,
    No springing up can make ;
Yet birds do sing and praise their king,
    And each one choose its mate.

MARCH.

Then in comes March, that noble arch,
    With wholesome Spring and air,
The child doth spring to years fifteen,
    With visage fine and fair ;
So do the flowers with softening showers,
    Aye spring up as we see ;
Yet, nevertheless, remember this,
    That one day we must die.

APRIL.

Then brave April doth sweetly smile,
    The flowers do fair appear,
The child is then become a man,
    To the age of twenty year.
If he be kind, and well inclin'd,
    And brought up at the school,
Then men may know if he foreshow,
    A wise man or a fool.

MAY.

Then cometh May, gallant and gay,
    When frequent flowers do thrive,
The child is then become a man,
    Of age twenty and five.
And for his life doth seek a wife,
    His days and years to spend ;
May He above send peace and love,
    And grace unto the end !

JUNE.

Then cometh June, with pleasant tune,
  When fields with flow'rs are clad,
And Phœbus bright is at his height,
  All creatures then are glad.
Then he appears of thirty years,
  With courage bold and stout ;
His nature so makes him to go,
  Of death he hath no doubt.

JULY.

Then July comes with his hot calms,
  And constant in his kind ;
The man doth thrive to thirty-five,
  And sober grows his mind ;
His children small do on him call,
  And breed him sturt and strife ;
His wife may die, and so may he
  Go seek another wife.

AUGUST.

Then August old, both stout and bold,
  When flow'rs do stoutly stand :
So man appears at forty years,
  With wisdom and command ;
And doth provide his house to guide,
  Children and familie ;
Yet do not miss t' remember this,
  That one day thou must die.

SEPTEMBER.

September then comes with his train,
  And makes the flowers to fade ;
Then man belyve is forty-five,
  Grave, constant, wise, and staid :

When he looks on, how youth is gone,
    And shall it no more see,
Then may he say, both night and day,
    Have mercy, Lord, on me!

OCTOBER.

October's blast comes in with boast,
    And makes the flow'rs to fall;
Then man appears at fifty years,
    Old age doth on him call;
The almond-tree doth flourish hie,
    And pale grows man we see;
Then it is time to use this line,
    Remember man to die.

NOVEMBER.

November air maketh fields bare,
    Of flowers, of grass, of corn;
Then man arrives at fifty-five,
    And sick both e'en and morn;
Loins, legs, and thighs, with sad disease,
    Make him to sigh and say,—
"Ah! Heaven on high have mind on me,
    And learn me how to die."

DECEMBER.

December fell, baith sharp and snell,
    Makes flowers creep in the ground;
Then man's threescore, both sick and sore,
    No soundness in him found.
His ears and een, and teeth of bane,
    All these now to him fail;
That he may say, both night and day,
    That death shall him assail.

The Garland of the Year.

JANUARY.

With his ice, and snow, and rime
Let bleak Winter sternly come !
There is not a sunnier clime
Than the love-lit winter home.

A. A. Watts.

JANUARY, whether bleak, rainy, and cold, or clear, frosty, and bracing, is one of the pleasantest times of the year. New Year's Day is usually considered, both in the old world and in the new, a most appropriate time for congratulations and good wishes, and for the giving of presents. It is, above all, the time for family gatherings and merry parties, as well as for out-door enjoyments, in the way of skating, curling, sliding, sleighing, or snow-balling.

The practice of making presents on New Year's Day was, no doubt, derived from the Romans. Suetonius and Tacitus both mention it. Claudius prohibited demanding presents, except on this day. Brand, in his *Popular Antiquities*, observes, on the authority of Bishop Stillingfleet, that the Saxons kept the festival of New Year with more than usual feasting and jollity, and with the presenting of New Year's gifts to each other. Fosbroke notices the continuation of the practice during the Middle Ages; and Ellis, in his additions to Brand, quotes Matthew Paris, to shew that Henry III. extorted New Year's gifts from his subjects. Those gifts presented by individuals to each other were suited to sex, rank, situation, and circumstances.

The custom of presenting New Year's gifts to the Sovereigns of England may be traced back to an early period. A manuscript roll of the public revenue of the fifth year of Edward VI. has an entry of rewards on New Year's Day to the king's officers and servants, amounting to £155 5s., and also of sums given to the servants of those who presented New Year's gifts to the king. A similar roll has been preserved of the reign of Philip and Mary. During the reign of Queen Elizabeth the custom of presenting New Year's gifts to the sovereign was carried to an extravagant height. The queen delighted in gorgeous dresses, in jewellery, in all kinds of ornaments for her person and palaces, and in purses filled with gold coin. The gifts regularly presented to her were of great value—those of the Earl of Leicester exceeded all others in costliness and elaborate workmanship.

In the reign of James I. money gifts were also presented; but the ornamental articles appear to have been fewer, and of less value. No rolls, or indeed any notices seem to have been preserved of New Year's gifts presented to Charles I.; though, probably, there were such. The custom, no doubt, ceased during the Commonwealth, and was not afterwards revived to any extent worthy of notice.

The practice of giving New Year presents, among relatives and friends in England, in America, and on the Continent of Europe is still continued; and it is kept up to a surprising extent in Paris, where the day is especially recognised from this circumstance as *le Jour d'Etrennes.* This is also the case in New York, in which city New Year's Day is entirely spent in giving and receiving visits of friends, in giving and receiving presents, and in all sorts of enjoyment and festivity.

By the early Romans this month was called *Januarius,* in honour of Janus, the deity supposed to preside over doors (Lat., *janua,* a door), who might very naturally be supposed to have something to do with the opening of the year. Janus is always represented as having two faces; one looking back on the old year, and the other looking forward to the new. In times of

war, among the Romans, his temple was always kept
open ; and for seven hundred years it was only shut
three times—one of these being in the reign of the Em-
peror Augustus, during which our Saviour was born.

Our Saxon ancestors called January *Wolf-monat,*
i.e., Wolf Month, because of the popular belief that the
wolves, that then infested the woods, were more daring
and voracious in this month than in any other—probably on account of its extreme cold in our northern
hemisphere. But, notwithstanding the severity of the
weather during this month, we may often notice, in
sheltered places, pretty flowers, few and far between,
both wild and garden, which enliven and brighten the
scene. The modest snow-drop, the yellow jasmine, the
blue periwinkle, the rugged gorse, the celandine, and
others ; with the evergreen ivy in full fruit, the ruddy
holly-berry, and the sacred mistletoe, cheer us with the
prospect of bright days yet to come, and remind us of
the goodness of an all-wise God. Of these, the snow-
drop comes first in order in our "Garland of the Year."
It has been suitably referred to by that admirable poet-
ess, Mrs. Browning, in the following beautiful lines :—

> " The poor, sad snow-drop,—growing between drifts,
> Mysterious medium 'twixt the plant and frost,
> So faint with Winter while so quick with Spring—
> So doubtful if to thaw itself away
> With that snow near it."

ORIGIN OF THE SNOW-DROP.

No fading flowers in Eden grew,
  Nor Autumn's withering spread
Among the trees a browner hue,
  To shew the leaves were dead ;
But through the groves and shady dells,
Waving their bright immortal bells,
Were amaranths and asphodels,
Undying in a place that knew
A golden age the whole year through.

But when the angel's fiery bands,
  Guarding the eastern gate,
Told of a broken law's commands,
  And agonies that came too late ;
With "longing, lingering" wish to stay,
And many a fond but vain delay
That could not wile her grief away,
Eve wandered aimless o'er a world
On which the wrath of God was hurled.

Then came the Spring's capricious smile,
  And Summer sunlight warmed the air,
And Autumn's riches served a while
  To hide the curse that lingered there ;
Till o'er the once untroubled sky
Quick driven clouds began to fly,
And moaning zephyrs ceased to sigh,
When Winter's storms in fury burst
Upon a world indeed accurst.

And when, at last, the driving snow—
  A strange, ill-omened sight—
Came whitening all the plains below,
  To trembling Eve it seemed—affright
With shivering cold and terror bowed,—
As if each fleecy vapour cloud
Were falling as a snowy shroud,
To form a close enwrapping pall
For Earth's untimeous funeral.

Then all her faith and gladness fled,
  And, nothing left but blank despair,
Eve madly wished she had been dead,
  Or never born a pilgrim there ;
But as she wept, an angel bent
His way adown the firmament,
And on a task of mercy sent
He raised her up, and bade her cheer
Her drooping heart, and banish fear :

And catching, as he gently spake,
  A flake of falling snow,
He breathed on it, and bade it take
  A form and bud and blow ;
And, ere the flake had reached the earth,
Eve smiled upon the beauteous birth,
That seemed, amid the general dearth
Of living things, a greater prize
Than all her flowers in Paradise.

"This is an earnest, Eve, to thee,"
The glorious angel said,
"That sun and Summer soon shall be ;
And, though the leaves seem dead.
Yet once again the smiling Spring,
With wooing winds shall swiftly bring
New life to every sleeping thing ;
Until they wake, and make the scene
Look fresh again and gaily green."

The angel's mission being ended,
  Up to Heaven he flew ;
But where he first descended,
  And where he bade the earth adieu,
A ring of snow-drops formed a posy
Of pallid flowers, whose leaves, unrosy,
Waved like a winged argosy,—
Whose climbing masts, above the sea,
Spread fluttering sail and streamer free.

And thus the snow-drop, like a bow
    That spans the cloudy sky,
Becomes a symbol whence we know
    That brighter days are nigh ;
That circling seasons, in a race,
That know no lagging, lingering pace,
Shall each the other nimbly chase,
Till Time's departing final day
Sweep snow-drops and the world away.

G. W.

THE SMALL FLOWER BURSTING ITS FROSTY PRISON.

ALL as the hungry winter-starvèd earth,
Where she by nature labours towards her birth,
Still as the day upon the dark world creeps,
One blossom forth after another peeps,
Till the small flower, whose root is now unbound,
Gets from the frosty prison of the ground,
Spreading the leaves unto the powerful noon,
Deck'd in fresh colours, smiles upon the Sun.

Drayton.

CROCUS.
FEBRUARY.

In the cheerless days of Springtime
　Arrayed in shining gold,
How welcome then the Crocus is,
　Just peeping from the mould ;
Or, robed in royal purple, shines
　In glory as a king.
And glows with lustrous beauty,
　The pride of early Spring.

FEBRUARY.

Ill fares the traveller now, and he that stalks,
In ponderous boots, beside his reeking team ;
The wain goes heavily, impeded sore
By congregated loads, adhering close
To the clogged wheels ; and, in its sluggish pace,
Noiseless, appears a moving hill of snow.

Cowper.

FEBRUARY, the shortest month of the year, is usually one of the most disagreeable and uncomfortable, owing to the extreme changeableness and uncertainty of the weather. To-day we may be in the depths of Winter ; to-morrow we may fancy ourselves in smiling Spring ; when, lo ! ere the day is gone, we may have stern Winter back again ! We may congratulate ourselves, however, that this joyless and miserable season will not be of long duration. The dry wind of March will speedily come, and clear away the dark clouds and the moisture, now too plentiful everywhere.

February takes its name from *Februa*, which signifies to expiate, or purify, from the fact that the religious expiation of the ancient Romans took place at the beginning of this month. Our Saxon ancestors gave it the name of *Sprout-kale*, from the fact of the sprouting of cabbage at this ungenial season. The name of *Sol-monat* was afterwards bestowed on it, in consequence of the return of the luminary of day from the low course in the heavens, which for some time he had been taking. The usual emblematical representation of February is a man in a sky-coloured dress, bearing in his hand the astronomical sign, Pisces.

"February fill-dyke" was a name formerly given to this month by our forefathers ; for, when the snow melted, the rivers overflowed, the dykes brimmed over, and long leagues of land were under water, which are now to be found drained, owing to improved agriculture. In this month no striking change has passed over the fields ; few flowers are to be noticed beyond those to be seen in the previous month. The verdure seems to belong more to the past than the present year ; yet, the primrose, the violet, the woodruff, the small white potentilla (often mistaken for wild strawberry), are to be met with in warm, or sheltered places ; and the simple daisy may be found embedded among the green grass in the fields.

WHEN THE SNOW HAS LEFT THE MOUNTAINS.

THE cock is crowing,
The stream is flowing,
The small birds twitter,
The lake doth glitter,
The green field sleeps in the sun ;
The oldest and the youngest
Are at work with the strongest ;
The cattle are grazing,
Their heads never raising ;
There are forty feeding like one.

Like an army defeated
The snow hath retreated,
And now doth fare ill
On the top of the bare hill ;
The ploughboy is whooping—anon—anon :
There's joy in the mountains,
There's life in the fountains ;
Small clouds are sailing,
Blue sky prevailing ;
The rain is over and gone.

Wordsworth.

AFTERNOON IN FEBRUARY.

THE day is ending,
The night is descending,
The marsh is frozen,
    The river is dead.

Through clouds, like ashes,
The red sun flashes
On village windows,
    That glimmer red.

The snow re-commences ;
The buried fences
Mark no longer
    The road o'er the plain ;

While through the meadows,
Like fearful shadows,
Slowly passes
    A funeral train.

The bell is pealing,
And every feeling
Within me responds
    To the dismal knell.

Shadows are trailing,
My heart is bewailing,
And tolling within
    Like a funeral bell.

Longfellow.

TO A CROCUS,

BLOSSOMING BENEATH A WALLFLOWER.

———◆◆———

WELCOME, wild harbinger of Spring,
  To this small nook of earth !
Feeling and fancy fondly cling
  Round thoughts which owe their birth
To thee, and to the humble spot
Where chance has fixed thy lowly lot.

To thee,—for thy rich golden bloom,
  Like Heaven's fair bow on high,
Portends, amid surrounding gloom,
  That brighter hours draw nigh,
When blossoms of more varied dyes
Shall ope their tints to warmer skies.

Yet not the lily, nor the rose,
  Though fairer far they be,
Can more delightful thoughts disclose
  Than I derive from thee ;
The eye their beauty may prefer,
The heart is thy interpreter !

Methinks in thy fair flower is seen,
  By those whose fancies roam,
An emblem of that leaf of green
  The faithful dove brought home ;
When o'er the world of waters dark
Were driven the inmates of the ark.

That leaf betokened freedom nigh
    To mournful captives there ;
Thy flower foretells a sunnier sky,
    And chides the dark despair
By Winter's chilling influence flung
O'er spirits sunk and nerves unstrung.

And sweetly has kind Nature's hand
    Assigned thy dwelling place
Beneath a flower whose blooms expand,
    With fond congenial grace,
On many a desolated pile,
Brightening decay with beauty's smile.

Thine is the flower of Hope, whose hue
    Is bright with coming joy ;
The wallflower's that of Faith, too true
    For ruin to destroy :
And where, O ! where should Hope up-spring
But under Faith's protecting wing.

Barton.

TO THE CROCUS.

Lowly, sprightly little flower !
    Herald of a brighter bloom,
Bursting in a sunny hour
    From thy Winter tomb.

Hues you bring, bright, gay, and tender,
  As if never to decay,
Fleeting in their varied splendour—
  Soon, alas ! it fades away.

Thus the hopes I long had cherished,
  Thus the friends I long had known,
One by one, like you have perished,
  Blighted—I must fade alone.

Patterson.

SPRING-TIME.

THOU wak'st again, O Earth,
  From Winter's sleep !
Bursting with voice of mirth
  From icy keep ;
And, laughing at the Sun,
Who hath their freedom won,
  Thy waters leap !
Thou wak'st again, O Earth !
  Freshly again,
And who by fireside hearth
  Now will remain ?

Come on thy rosy hours,—
Come on the buds and flowers,
As when in Eden's bowers
    Spring first did reign !
Birds on thy breezes chime,
Blithe as in that matin-time
    Their choiring begun :
Earth, thou hast many a prime,—
    Man hath but one.

Thou wak'st again, O Earth !
    Freshly and new,
As when at Spring's first birth
    First flowerets grew.
Heart ! that to Earth doth cling
While boughs are blossoming,
    Why wake not too ?
Long thou in sloth hast lain,
Listing to Love's soft strain—
    Wilt thou sleep on ?
Playing, thou sluggard heart,
In life no manly part,
    Though youth be gone.
Wake ! 'tis Spring's quickening breath
    Now o'er thee blown :
Wake thee ! and e'er in death
Pulseless thou slumbereth,
Pluck but from Glory's wreath
    One leaf alone !

                                Hoffman.

PRIMROSE
MARCH.

Fairest child of
        the opening year!
Delicate primrose,
    how welcome, how dear
The sight once more
    of thy star-like flower—
The joy of many
            a happy hour;
When children gay
        sweet posies bring,
And joy in the beauty
        of welcome Spring,

MARCH.

How pleasant 'tis to mark the labouring plough
Traverse the field, and leave a sable track,
While merrily behind the driver stalks,
Whistling, in thoughtless vacancy of mind.

D. M. Moir.

WELCOME, March! Adieu, rude Winter, weird and wild. "Come, gentle Spring, ethereal mildness, come! Come, ye vernal breezes and refreshing showers, come, and welcome in the lengthening days!" Nature begins to look gay, and to don her robes of brightest green. Spring is come at last, and March is here. Now, on the very first of the month there is a dry, chill air, with breaks of bright sunshine here and there over the landscape. The clouds move faster. The paths which in February were wet and sloppy, are now solid and dry, and the East wind sends clouds of dust along the roads. 'Tis glorious weather for the husbandman, who loses not a moment in digging, ploughing, and harrowing the earth—in planting and sowing ; for he looks for a rich return in Autumn. He prizes the dusty day, for he bears in mind that "A peck of March dust is worth a king's ransom."

March, which, with the Ancients, was the first month in the year, takes its name from Mars, the god of war, as it was the month in which wars and expeditions were usually undertaken, both by the Romans and the Goths. It was called by the Romans, *Martius.* Our Saxon forefathers termed it *Lenet-monat, i.e.,* Length Month, in reference to the lengthening of the days at this season, which is the origin also of the term Lent. March is represented as a man of fierce aspect, with blossoms and a basket of seeds on his left arm, and, in his right hand, the sign of Aries, or the Ram.

With March we now have a large selection of wild flowers from which to form a bouquet. Chief among all are the fragrant primroses and the sweet-scented wood violets (purple and white), which impregnate the very winds with their fragrance. The daisy (or day's eye, as the Saxons called it), and the marsh marigold brighten the fields. The hardy wind-flower or wood-anemone (some snow-white, and others tinged with a delicate pink or blue), the germander-speedwell and ivy-leafed veronica, with others, may be seen in the woods, or on the bare hill-side, soon, however, to pass away, and be replaced by other flowers, bearing the more brilliant tints of Summer.

JOLLY SPRING, AGED WINTER.

EARTH is now green, and heaven is blue ;
Lively Spring, which makes all new,
   Jolly Spring doth enter ;
Sweet young sunbeams do subdue
   Angry, aged Winter.
Winds are mild, and seas are calm,
Every meadow flows with balm,
   The earth wears all her riches ;
Harmonious birds sing such a psalm
   As ear and heart bewitches.

Sir J. Davies.

THE VIOLET.

SWEET violets, Love's paradise, that spread
   Your gracious odours, which you couchèd bear
      Within your palie faces,
Upon the gentle wing of some calm-breathing wind,
      That plays amidst the plain,
If by the favour of propitious stars you gain

Such grace as in my ladie's bosom place to find,
        Be proud to touch those places !
Your honours of the flowrie meads I pray,
    You pretty daughters of the earth and sun.

                                        Raleigh.

TO SPRING.

HAIL, verdant Spring ! we welcome thee !
    With breath so fresh, and looks so bright !
Again Sol shines with brilliant ray,
    And Cynthia's charms illume the night.

The fragrant primrose lifts its head,
    The russet mould gives forth gay flowers,
Which, o'er the fields and gardens spread,
    Seem brought to life by gentle showers.

On every tree young leaves shoot forth,
    In varying shades of brightest green,
And birds, with notes of gayest mirth,
    And sweetest sounds, enjoy the scene.

Each little songster of the grove
    Now fondly searches for a mate,
And thus seems anxious so to prove
    Its love for " the united state."

Thus as sweet Spring is bright and fair,
  Should youthful hearts be blythe and gay ;
Enlivened as by Spring-time's air
  In sweet contentment's genial ray.

Life is but short ; then let's enjoy
  The gifts kind Providence has sent ;
And let not discord's voice destroy
  The bliss that's in a life well spent.

The Editor.

TO A PRIMROSE.

Welcome, pale primrose, starting up between
  Dead matted leaves of oak and ash, that strew
  The every lawn, the wood and meadow through,
'Mid creeping moss and ivy's darker green.
  How much thy presence beautifies the ground !
How sweet thy modest unaffected pride
Glows on the sunny bank and wood's warm side !
  And where thy fairy flowers in groups are found,
The school-boy roams enchantedly along,
  Plucking the fairest with a rude delight ;
While the meek shepherd stays his simple song
  To gaze a moment on the pleasing sight,
O'erjoyed to see the flowers that truely bring
The welcome news of sweet returning Spring.

Clare.

THE PRIMROSE.

THE Sun declines ; his parting ray
Shall bear the cheerful light away,
   And on the landscape close ;
Then will I seek the lonely vale,
Where sober evening's primrose pale
   To greet the night-star blows.

Soft, melancholy bloom, to thee
I turn with conscious sympathy !
   Like thee my hour is come,
When lengthening shadows slowly fade,
Till, lost in universal shade,
   They sink beneath the tomb.

By thee I'll sit and inly muse ;
What are the charms in life we lose
   When time demands our breath ?
Alas ! the load of lengthen'd age
Has little can our wish engage,
   Or point the shaft of death.

No ; 'tis alone the pang to part
With those we love that rends the heart ;
   That agony to save,
Some nameless cause in nature strives.
Like thee, in shades our hope revives,
   And blossoms in the grave.

               Mrs. Hunter.

VIOLETS.—A SONNET.

BEAUTIFUL are you in your lowliness ;
  Bright in your hues, delicious in your scent,
  Lovely your modest blossoms, downward bent,
As shrinking from our gaze, yet prompt to bless
The passer-by with fragrance, and express
  How gracefully, though mutely eloquent,
  Are unobtrusive worth and meek content,
Rejoicing in their own obscure recess.
  Delightful flowerets ! at the voice of Spring
Your buds unfolded to its sunbeams bright ;
And, though your blossoms soon shall fade from sight,
  Above your lonely birth-place birds shall sing,
  And from your clustering leaves the glow-worm fling
The emerald glory of its earth-born light.

Barton.

THE VIOLET.

SWEET lowly plant, once more I bend
  To hail thy presence here,
Like a belov'd returning friend,
  From absence doubly dear.

Wert thou for ever in our sight,
　　Might we not love thee less ?
But now thou bringest new delight,
　　Thou still hast power to bless.

Still doth thy lovely presence bring
　　Of Spring-time joys a dream,
When life was in its sunny spring—
　　A fair, unrippled stream.

And still thine exquisite perfume
　　Is precious as of old,
And still thy modest, tender bloom
　　It joys me to behold.

It joys and cheers whene'er I see
　　Pain on earth's meek ones press,
To think the storm that rends the tree
　　Scathes not thy lowliness.

And thus may human weakness find,
　　E'en in thy lowly flower,
An image cheering to the mind
　　In many a trying hour.

　　　　　　　　　　　　　　Anon.

DAFFODIL.
APRIL.

Daffodils, in beauty rich,
  Robe the copse in cloth of gold,
Teach a useful lesson, which
  'T is a pleasure to unfold:
Telling that your pretty flower,
  Now in verdure freely springing.
Symbol is of nature's power,
  Beauty always to us bringing;
And of an Almighty care
  Spreading blessings everywhere.

APRIL.

Hail, April! true Medea of the year,
That makest all things young and fresh appear.
What praise, what thanks, what commendations due,
For all thy pearly drops of morning dew?
When we despair, thy seasonable showers
Comfort the corn, and cheer the drooping flowers;
As if thy charity could not but impart
A shower of tears to see us out of heart.
Sweet, I have penned thy praise, and here I bring it;
In confidence the birds themselves will sing it.

APRIL, month of showers and sunshine, clear and
cloudy skies, greenness and bareness, heavy hail
and blooming blossoms, we welcome thee! The
earth now begins, before the strengthening sun, to expand the flowers and plants, which adorn its surface
with every species of coloured beauty. Smiling Spring
is now in full bloom. The hedges and trees begin to
bud, and make a show of green. The bees are busy
among the bloom. The migratory birds have returned,

and the songsters of the grove now carol their sweetest lays. The cuckoo, the swallow, and the thrush, may be heard, and "the lark at Heaven's gate sings." The labours of the field are rapidly progressing ; and sunshine and showers, alternately, nourish the grain in the fields, as well as the flowers in the garden.

April is supposed to take its name from the Latin, *Aprilis*, or *aperio*, I open, indicating the opening in Spring of the buds of the trees and flowers. It is probable that the name *Aprilis* was *Aphrilis*, founded on the Greek name of the goddess Venus. Our Anglo-Saxon forefathers called this month *Oster-monat*, or *Easter-monat*, from the feast of their goddess Eastre, as some authorities state. Others say that the name was given on account of the East winds that usually prevail at this season. The term *Easter*, given to our Christian festival, may come from the same derivation.

Among the many pretty flowers that may now be seen are the showy daffodil, the bluebell, the wild anemone or wind-flower, the modest daisy (or *petite marguerite*, as the French call it), the gaudy dandelion, the lesser celandine, the oxlips and cowslips, and the violets and primroses, all of which are now blooming in abundance. As the month advances, and we approach its end, the brilliancy of the bloom increases, till we enter May, which has always been considered the most lovely month of the year.

FLOWERS O' THE SPRING.

DAFFODILS,
That come before the swallow dares, and take
The winds of March with beauty ; violets, dim,
But sweeter than the lids of Juno's eyes,
Or Cytherea's breath ; pale primroses,
That die unmarried ere they can behold
Bright Phœbus in his strength—a malady
Most incident to maids ; bold oxlips, and
The crown imperial ; lilies of all kinds—
The flowers-de-luce being one !   O ! these I lack
To make you garlands of.

Shakespeare.

DAFFODILS.

FAIR daffodils, we weep to see
You haste away so soon ;
As yet the early-rising sun
Has not attain'd his noon.
Stay, stay,

Until the hasting day
        Has run
But to the even-song ;
And, having prayed together, we
        Will go with you along.

We have short time to stay as you,
        We have as short a Spring ;
As quick a growth to meet decay
        As you, or any thing.
                We die
        As your hours do, and dry
                Away,
        Like to the Summer's rain,
Or, as the pearls of morning's dew,
        Ne'er to be found again.

                                Herrick.

APRIL.

CAPRICIOUS month of smiles and tears,
    There's beauty in thy varied reign !
Emblem of being's hopes and fears,
    Its hours of joy and days of pain.
A false, inconstant scene is thine,
    Changeful with light and shadow deep,
Oft-times thy clouds with pure sunshine
    Are painted ; then in gloom they sleep.

Yet is there gladness in thy hours,
   Frail courier of a brighter scene,
Thou fragrant guide to buds and flowers,
   To meadows fresh and pastures green !
For, as thy days grow few and brief,
   The radiant looks of Spring appear,
With swelling glow and opening leaf,
   To deck the morning of the year.

Yes, though thy life is checkered oft
   With drilling showers of sorrowing rain,
Yet balmy airs and breezes soft
   Are lingering richly in thy train ;
And for thy eddying gusts will come
   The lay of the rejoicing bird, —
That tries his new and brightening plume, —
   'Mid the void sky's recesses heard.

And soon the many clouds, that hang
   Their solemn drapery o'er the sky,
Will pass in shadowy folds away.
   Lo ! mark them now !—they break—they fly !
And over earth, in one broad smile,
   Looks forth the glorious eye of day ;
While hill, and vale, and ocean-isle
   Are laughing in the breath of May.

Type of existence ! mayst thou be
   The emblem of the Christian's race,
Through all whose trials we may see
   The sunshine of undying grace :

The calm and Heaven-enkindled eye,
The faith that mounts on ardent wing,
That looks beyond the o'er-arching sky
To Heaven's undimmed and golden spring.
Anon.

THE SOLITARY MOUNTAIN DAISY.

ON TURNING IT DOWN WITH THE PLOUGH.

WEE, modest, crimson-tippèd flow'r,
Thou's met me in an evil hour,
For I maun crush amang the stoure
Thy slender stem ;
To spare thee now is past my pow'r,
Thou bonnie gem !

Alas ! it's no thy neebor sweet,
The bonnie lark, companion meet !
Bending thee 'mang the dewy weet,
Wi' spreckled breast !
When upward springing, blythe, to greet
The purpling East.

Cauld blew the bitter, biting North
Upon thy early, humble birth,
Yet cheerfully thou glinted forth
Amid the storm, —
Scarce reared above the parent earth
Thy tender form.

The flaunting flow'rs our gardens yield,
High shelt'ring woods and wa's maun shield ;
But, thou, beneath the random bield
        O' clod or stane,
Adorns the histie stibble-field
          Unseen, alane.

There, in thy scanty mantle clad,
Thy snawy bosom sun-ward spread,
Thou lifts thy unassuming head
        In humble guise ;
But, now, the share uptears thy bed,
          And low thou lies !

Such is the fate of artless maid,
Sweet flow'ret of the rural shade !
By love's simplicity betrayed,
        And guileless trust,
Till she, like thee, all soil'd is laid
          Low i' the dust.

Such is the fate of simple Bard,
On Life's rough ocean luckless starr'd !
Unskilful he to note the card
        Of prudent lore,
Till billows rage, and gales blow hard,
          And whelm him o'er !

Such fate to suffering worth is given,
Who long with wants and woes has striv'n,

By human pride or cunning driv'n
                    To mis'ry's brink,
Till, wrench'd of ev'ry stay but Heav'n,
                    He, ruined, sink.

Ev'n thou who mourn'st the daisy's fate,
That fate is thine—no distant date ;
Stern Ruin's ploughshare drives, elate,
                    Full on thy bloom,
Till, crush'd beneath the furrow's weight,
                    Shall be thy doom !

Burns.

THE EYE OF THE DAY.

THESE flow'rés, white and red,
Such that men callen daisies in our town ;
To them have I so great affection,
As I said erst, when comen is the May,
That in my bed there daweth me no day
That I n'am up and walking in the mead
To see this flow'r against the sunné spread,
When it upriseth early by the morrow ;
That blissful sight softeneth all my sorrow,
So glad am I when that I have presénce
Of it, to doen it all revérence.

Chaucer.

HAWTHORN. MAY.

Hawthorn blossoms,
      bright and fair,
Summer's sun and
      scented air;
Through the verdure
      flow'rets come,
In the shade
      the insects hum,—
Song-birds singing
      on the spray:
Welcome, gladly,
      welcome, May!

MAY.

Hail, beauteous May ! thou dost inspire
Mirth and youth and warm desire ;
Woods and groves are of thy dressing,
Hill and dale doth boast thy blessing.
Thus we salute thee with our early song,
And welcome thee, and wish thee long.

Milton.

MAY ! charming May ! gay in all thy beauty, with the health, wealth, joyfulness, and youth of the year ; fresh with the fragrance of the hawthorn, the violet and primrose, and other vernal shrubs and flowers. " Lo ! the Winter is past, the rain is over and gone ; the flowers appear on the earth, the time of the singing of birds is come, and the voice of the turtle is heard in our land." There is now a rich greenness in the leaf of the young corn and grass, and in the young leaves and buds of the trees, bursting into youthful beauty. May is described as " a beautiful maiden, clothed in sunshine, and scattering flowers

on the earth, while she dances to the music of birds
and brooks." The birds now sing their enraptured
songs of love. The bees hover from flower to flower.
The butterflies, in the most brilliant colours, are dart-
ing zig-zag in every direction, in the green lanes and
fields. Everything seems clothed in beauty.

May is said by some authorities to have received
the name in honour of Maia, the mother, by Jupiter,
of the god Hermes or Mercury ; but others state that
the name was assigned to it by Romulus, the founder
of Rome, in honour of his nobles or senators, who were
called *Majories,* or, *Maiores.* The Anglo-Saxons
called it *Tri-milchi,* with reference, probably, to the
improved condition of cattle, from the benefit of the
Spring herbage as food.

In this month the flowers are abundant on every
side. The dazzling white of the daisies, the glittering
gold of the buttercups, the fragrant lily of the valley,
the sweet woodruff, the red and white campions, the
greater and the lesser stitchwort, the various orchises,
and the wild geranium, charm the eye, and the singing
of the birds delights the ear, as we walk abroad in the
green lanes or meadows ; while the beauty of budding
foliage on the trees and hedges, and the fragrance of
the hawthorn make us grateful to an all-wise Providence for the varying beauty of the seasons in this
happy land, in which our lot is cast.

SPRING.

Now the lusty Spring is seen ;
  Golden yellow, gaudy blue,
  Daintily invite the view.
Everywhere, on every green,
Roses, blushing as they blow,
  And enticing men to pull ;
Lilies, whiter than the snow,
  Woodbines of sweet honey full ;
All Love's emblems, and all cry—
"Ladies, if not plucked, we die."

Beaumont and Fletcher.

TO BLOSSOMS.

Fair pledges of a fruitful tree,
  Why do ye fall so fast ?
  Your date is not so past

But you may stay yet here awhile,
To blush and gently smile,
   And go at last.

What ! were ye born to be
   An hour or half's delight,
   And so to bid good-night ?
'Tis pity Nature brought ye forth
Merely to show your worth,
   And lose you quite.

But you are lovely leaves, where we
   May read how soon things have
   Their end, though ne'er so brave ;
And, after they have shown their pride,
Like you, awhile, they glide
   Into the grave.

Herrick.

SONG TO MAY.

MAY ! Queen of blossoms
   And fulfilling flowers,
With what pretty music
   Shall we charm the hours ?
Wilt thou have pipe and reed,
Blown in the open mead,
Or to the lute give heed
   In the green bowers ?

Thou hast no need of us,
    Or pipe, or wire,
That hast the golden bee,
    Ripened with fire ;
And many thousands more
Songsters that thee adore,
Filling earth's grassy floor
    With new desire.

Thou hast thy mighty herds,
    Tame and free livers ;
Doubt not, thy music too
    In the deep rivers ;
And the whole plumy flight,
Warbling the day and night,
Up at the gates of light :—
    See ! the lark quivers !

When, with the jacinth,
    Coy fountains are tressed,
And for the mournful bird
    Green woods are dressed,
That did for Tereus pine,
Then shall our songs be thine,
To whom our hearts incline :
    May, be thou blessed !

                                Lord Thurlow.

THE COWSLIP.

UNFOLDING to the breeze of May,
The cowslip greets the vernal ray ;
The topaz and the ruby gem
Her blossoms' simple diadem ;
And, as the dew-drops gently fall,
They tip with pearls her coronal.
In princely halls and courts of kings
Its lustrous ray the diamond flings ;
Yet few of those who see its beam,
Amid the torches' dazzling gleam,
As bright as though a meteor shone,
Can call the costly prize their own.

And gems, of every form and hue,
Are glittering here in morning dew ;
Jewels that all alike may share
As freely as the common air ;
No niggard hand, no jealous eye,
Protects them from the passer-by.

Man to his brother shuts his heart,
And Science acts a miser's part ;
But Nature, with a liberal hand,
Flings wide her stores o'er sea and land.
If gold she gives, not single grains
Are scattered far across the plains,

But, lo ! the desert streams are rolled
O'er precious beds of virgin gold.
If flowers she offers, wreaths are given,
As countless as the stars of Heaven !
If music—'tis no feeble note
She bids along the valley float ;
Ten thousand nameless melodies,
In one full chorus, swell the breeze.

Oh, Art is but a scanty rill,
That genial seasons scarcely fill,
But Nature needs no tides' return
To fill afresh her flowing urn ;
She gathers all her rich supplies
Where never-failing fountains rise.

Anon.

MAY.

Now, while the birds thus sing a joyous song,
  And while the young lambs bound
  As to the tabor's sound,
To me alone there came a thought of grief ;
A timely utterance gave that thought relief,
  And I again am strong ;
The cataracts blow their trumpets from the steep,
  No more shall grief of mine the season wrong ;
  I hear the echoes through the mountains throng,

The winds come to me from the fields of sleep,
    And all the earth is gay ;
        Land and sea
    Give themselves up to jollity,
        And with the heart of May
    Doth every beast keep holiday.
        Thou child of joy,
Shout round me, let me hear thy shouts, thou happy
        Shepherd Boy !

Ye blessed creatures, I have heard the call
    Ye to each other make ; I see
The heavens laugh with you in your jubilee ;
    My heart is at your festival,
    My head hath its coronal,
The fulness of your bliss I feel—I feel it all.
    Oh, evil day ! if I were sullen,
    While the Earth herself is adorning
        This sweet May-morning,
    And the children are pulling,
            On every side,
    In a thousand valleys, far and wide,
Fresh flowers ; while the sun shines warm,
And the babe leaps up on his mother's arm.
                                Wordsworth.

OX-EYE DAISY

JUNE

On the fragrant mead,
   Among the new-mown hay,
We see the daisies tall,
   That bloomed but yesterday,
Now with'ring with the grass
   Their transient glory gone,
And in their lesson see
   The fate of everyone.

JUNE.

'Tis June, 'tis merry, smiling June,
  'Tis blushing Summer now ;
The rose is red, the bloom is dead,
  The fruit is on the bough.

Eliza Cook.

JUNE, sweet June ! month of roses and lovely
flowers ! Adieu to Spring—welcome Summer !
Adieu to the modest snow-drops, primroses, and
violets ! Adieu to the pale green leaflets of Spring !
They are gone—gone to be speedily replaced by the
more brilliant flowers and leaves of Summer, now in
full glory. The forest trees put forth their blossoms.
The earth is carpeted with flowers, and the air is filled
with their perfume. The flower-garden is now in its
splendour : roses of all tints, from deepest crimson to
palest pink, and from brightest yellow to palest white ;
besides stocks, geraniums, lilies, speedwells, jasmines,

rockets, poppies, pinks, lupines, mignionette, and hundreds of others. In the fields and country lanes, we are delighted with the sight and perfume of the field-peas and beans, the red and white clover, the young corn, bursting into ear, the delicate wild rose, with flowers of varying tints, the luscious honeysuckle, the snowy-flowered elder, the foxglove, the meadow-sweet, and the exquisite feathery grasses of the field, which rival the beauty of the plumes of the ostrich.

It is supposed by some that June took its name from Juno, the wife of Jupiter, "king of gods and men." Others say it is derived *à junioribus*, from young persons, who always claim this month as their own. The ancient Romans considered that June was the most propitious season of the year for contracting matrimonial engagements, and, particularly so, at the full of the moon, and that the month of May was especially to be avoided.

The out-door labours in the field are specially pleasant now. The happy haymakers in the meadow appear as if their work was only pleasure. June is sometimes showery ; but sunshine quickly comes, and then we have, in all its splendour, the rainbow—"triumphal arch, which fill'st the sky when storms prepare to part." But sunshine and showers alternating only enhance the beauty of the season, which may be considered one of the most enjoyable times of the year.

FIELD FLOWERS.

YE field flowers! the gardens eclipse you, 'tis true ;
Yet, wildings of Nature, I doat upon you ;
 For ye waft me to Summers of old,
When the earth teemed around me with fairy delight,
And when daisies and buttercups gladdened my sight,
 Like treasures of silver and gold.

Not a pastoral song has a pleasanter tune
Than ye speak to my heart, little wildings of June ;
 Of ruinous castles ye tell,
Where I thought it delightful your beauties to find,
When the magic of Nature first breathed on my mind,
 And your blossoms were part of her spell.

Even now what affection the violet awakes ;
What loved little islands, twice seen in their lakes,
 Can the wild water-lily restore ;
What landscapes I read in the primrose's looks,
And what pictures of pebbled and minnowy brooks
 In the vetches that tangled their shore !

Earth's cultureless buds, to my heart ye were dear,
Ere the fever of passion, or ague of fear,
  Had scathed my existence's bloom ;
Once I welcomed you more in Life's passionless stage,
With the visions of youth to revisit my age,
  And I wish you to grow on my tomb.

Campbell.

JUNE.

Now comes the rosy June ! and blue-eyed hours,
  With song of birds and stir of leaves and wings,
  And run of rills, and bubble of bright springs,
And hourly burst of pretty buds to flowers ;
With buzz of happy bees in violet bowers,
  And gushing lay of the loud lark, who sings
  High in the silent air, and sleeks his wings
In frequent sheddings of the flying showers ;
  With plunge of struggling sheep in plashy floods,
And timid bleat of shorn and shivering lamb,
Answer'd in far-off faintness by its dam ;
  With cuckoo's call from green depths of old woods,
And hum of many sounds, making one voice
That sweetens the smooth air with a melodious noise.

Waller.

THE DAISY.

TRAMPLED underfoot,
The daisy lives, and strikes its little root
Into the lap of Time ; centuries may come,
And pass away into the silent tomb,
And still the child, hid in the womb of Time,
Shall smile and pluck them !   When this simple rhyme
Shall be forgotten, like a church-yard stone,
Or lingering lie, unnoticed and alone ;
When eighteen hundred years, our common date,
Grow many thousands in their marching state,
Ay, still the child, with pleasure in his eye,
Shall cry, " The daisy ! "—a familiar cry—
And run to pluck it in the self-same state ;
And, like a child himself, when all was new,
Might smile with wonder, and take notice too !
Its little golden bosom, filled with snow,
Might win e'en Eve to stoop adown and shew
Her partner, Adam, in the silken grass,
The little gem, that smiled where pleasure was,
And, loving Eve, from Eden followed ill,
And bloomed with sorrow, and lives smiling still ;
As once in Eden, under Heaven's breath,
So now on Earth, and on the lap of death,
It smiles for ever.

Clare.

SONNET.

WRITTEN AT THE CLOSE OF SPRING.

THE garlands fade that Spring so lately wove ;
    Each simple flower, which she had nurs'd in dew,
Anemones, that spangled every grove,
    The primrose wan, and harebell, mildly blue.
No more shall violets linger in the dell,
    Or purple orchis variegate the plain,
Till Spring again shall call forth every bell,
    And dress with humid hands her wreaths again.

Ah, poor humanity ! so frail, so fair,
    Are the fond visions of thy early day,
Till tyrant passion and corrosive care
    Bid all thy fairy colours fade away !
Another May new buds and flowers shall bring :
Ah ! why has happiness no second spring?

Charlotte Smith.

THE MOSS ROSE.

THE Angel of the flowers, one day,
Beneath a rose-tree sleeping lay ;
That spirit to whose charge 'tis given
To bathe young buds in dews of Heaven.
Awaking from his light repose,
The angel whispered to the rose :

" O, fondest object of my care,
Still fairest found where all are fair,
For the sweet shade thou giv'st to me,
Ask what thou wilt—'tis granted thee."
" Then," said the rose, with deepened glow,
" On me another grace bestow ! "
The spirit paused in silent thought, —
What grace was there that flower had not ?
'T was but a moment ; o'er the rose
A veil of moss the angel throws ;
And, robed in Nature's simplest weed,
Could there a flower that rose exceed ?

TO A DAISY.

BRIGHT flower, whose home is everywhere !
A pilgrim bold, in Nature's care,
And oft, the long year through, the heir
        Of joy or sorrow ;
Methinks that there abides in thee
Some concord with humanity,
Given to no other flower I see
        The forest thorough !

And wherefore ?   Man is soon depressed ;
A thoughtless thing who, once unblest,
Does little on his memory rest,
        Or on his reason ;

But thou wouldst teach him how to find
A shelter under every wind,
A hope for times that are unkind,
　　And every season.

Wordsworth.

THE ROSE.

(FROM CAMOENS.)

——♦♦——

JUST like Love is yonder rose :—
Heavenly fragrance round it throws,
Yet tears its dewy leaves disclose,
And, in the midst of briars it blows, —
　　　　　　Just like Love !

Culled to bloom upon the breast,
Since rough thorns the stem invest,
They must be gathered with the rest,
And with it to the heart be prest,—
　　　　　　Just like Love !

And when rude hands the twin-buds sever,
They die, and they shall blossom never ;
Yet the thorns be sharp as ever ;—
　　　　　　Just like Love !

Anon.

CONVOLVULUS

JULY

In the dew of Summer morn,

Are thy varied blossoms born

Born to meet the Sun's first ray,

Soon, too soon, to pass away.

Look upon them, for they say

"Do your duty while you may,

Lest you die, your work undone

And may see no morrow's sun."

JULY.

'Tis Summer—joyous Summer-time !
In noisy towns no more abide ;
The earth is full of radiant things,
Of gleaming flowers and glancing wings—
Beauty and joy on every side.

Mary Howitt.

SWEET Summer has now attained her perfection. 'Tis burning July, month of heat and sunshine, of azure skies and dusty roads, of ripened hay and ripening corn. The fields are nigh white for harvest. The glowing landscape shows a picture of brightness and warmth. At noon-time we gladly seek the pleasant shade of the trees, so richly clothed with bright foliage. We sigh for the cooling breeze, or freshening shower of rain. In the flower-garden the show on the beds is gorgeous ; the lilies "shine in glory as a king ;" yet the perfume of the roses and other flowers is fairly rivalled in the fields by the fragrance of the new-mown hay. The busy haymakers are at work, piling it into hay-cocks, or carrying it homewards to the stack-yard. In the city, residence is almost unendurable, and all who can leave now gladly seek the fresh, cool air of the country, or the breeze at the seaside.

The ancient Romans called this month *Quintilis.* It was the fifth month of the Roman year, and had originally thirty-six days. It was the natal month of Julius Cæsar, who, in reforming the Calendar, allotted to it thirty-one days. It was named *July* by Mark Antony, in compliment to him. Our Saxon forefathers termed it *Hey-monat*, because they then made their hay harvest ; and also *Maed-monat*, from the meads being then in full bloom.

The wild flowers of Spring have entirely disappeared. Climbing plants festoon the hedges. The wild hop, the bryony, the large white convolvulus, and others, deck the bushes with varied beauty, and breathe the Summer's sweetness. In the fields, the scarlet poppy, the blue-bottle, the marigold, and the dog-daisy, may be seen in abundance. On the roadsides and ditches, among beautiful ferns, may be seen the tall foxglove, the musk-thistle, the wild thyme, and hosts of others, which brighten the way of the weary foot passenger along the dusty road.

JULY.

Loud is the Summer's busy song,
The smallest breeze can find a tongue ;
While insects of each tiny size
Grow teasing with their melodies,
Till noon burns with its blistering breath
Around, and day lies still as death.

The busy noise of man and brute
Is, on a sudden, lost and mute ;
Even the brook, that leaps along,
Seems weary of its bubbling song,
And, so soft its waters creep,
Tired silence sinks in sounder sleep.

The cricket on its bank is dumb,
The very flies forget to hum ;
And, save the waggon rocking round,
The landscape sleeps without a sound ;
The breeze is stopped, the lazy bough
Hath not a leaf that danceth now ;

The taller grass upon the hill
And spiders' threads are standing still ;

The feathers, dropped from moor-hen's wing,
Which to the water's surface cling,
Are steadfast, and as heavy seem
As stones beneath them in the stream ;

Hawkweed and groundsel's fanny downs
Unruffled keep their seedy crowns ;
And in the over-heated air
Not one light thing is floating there,
Save that, to the earnest eye,
The restless heat seems twittering by.

Noon swoons beneath the heat it made,
And flowers e'en within the shade,
Until the sun slopes in the West.
Like weary traveller, glad to rest
On pillowed clouds of many hues,
Then Nature's voice its joy renews ;

And checkered field and grassy plain
Hum, with their Summer-songs again,
A requiem to the day's decline,
Whose setting sunbeams coolly shine,
As welcome to day's feeble powers
As falling dews to thirsty flowers.

John Clare.

SUMMER MORN.

WITH quickened step
Brown night retires ; young day pours in apace,
And opens all the lawny prospect wide.
The dripping rock, the mountain's misty top
Swell on the sight, and brighten with the dawn.
Blue, through the dusk, the smoking currents shine,
And from the bladed field the fearful hare
Limps awkward ; while, along the forest glade,
The wild deer trip, and, often turning, gaze
At early passenger.   Music awakes
The native voice of undissembled joy ;
And, thick around the woodland, hymns arise.
Roused by the cock, the soon-clad shepherd leaves
His mossy cottage, where with peace he dwells ;
And from the crowded fold, in order, drives
His flock, to taste the verdure of the morn.

James Thomson.

SUMMER EVE.

Low walks the sun, and broadens by degrees,
Just o'er the verge of day.   The shifting clouds
Assembled gay, a richly gorgeous train,
In all their pomp attend his setting throne.
Air, earth, and ocean smile immense, and now,

As if his weary chariot sought the bowers
Of Amphitrité and her tending nymphs
(So Grecian fable sung), he dips his orb,
Now half immersed, and now a golden curve,
Gives one bright glance, then total disappears.
Confessed, from yonder slow-extinguished clouds,
All ether softening, sober evening takes
Her wonted station in the middle air,
A thousand shadows at her beck.

James Thomson.

THE LILY AND THE ROSE.

THE Nymph must lose her female friend
    If more admired than she ;—
But where will fierce contention end
    If flowers can disagree ?

Within the garden's peaceful scene
    Appeared two lovely foes,
Aspiring to the rank of queen,
    The Lily and the Rose.

The Rose soon reddened into rage,
    And, swelling with disdain,
Appealed to many a poet's page
    To prove her right to reign.

The Lily's height bespoke command—
    A fair imperial flower ;

She seemed designed for Flora's hand,
  The sceptre of her power.

This civil bickering and debate
  The goddess chanced to hear,
And flew to save, ere yet too late,
  The pride of the parterre.

"Yours is," she said, "the nobler hue,
  And yours the statelier mien,
And, till a third surpasses you,
  Let each be deemed a queen!"

Thus soothed and reconciled, each seeks
  The fairest British fair ;
The seat of empire is her cheeks,
  They reign united there.

Cowper.

THE LILY.

THERE is a pale and modest flower,
  In garb of green array'd,
That decks the rustic maiden's bower,
  And blossoms in the glade ;
Though other flowers around me bloom,
  In gaudy splendour drest,
Filling the air with rich perfume,
  I love the lily best.

I see the tulip's gorgeous hue,
   And sun-flower's crown of gold ;
I see the rose, and woodbine too,
   Their scented leaves unfold ;
Though they adorn the gay parterre,
   I love them not so well
As the drooping lily, frail and fair.
   That grows in shady dell.   *Anon.*

THE LESSON OF A ROSE.

AH ! see, whose fayre thing dost faine to see,
In springing flowre the image of the day !
Ah ! see the virgin rose, how sweetly shee
Doth first peepe forth with bashful modestee,
That fairer seems the lesse ye see her may !
Lo ! see soone after how, more bold and free,
Her barèd bosome she doth broad display :
Lo ! see soone after how she fades and falls away !

So passeth, in the passing of a day
Of mortal life, the leafe, the bud, the flowre
No more doth flourish, after first decay,
That earst was sought to deck both bed and bowre
Of many a lady and many a paramoure !
Gather therefore the rose whilest yet is prime,
For soon comes age that will her pride deflowre ;
Gather the rose of love whilest yet is time,
Whilest loveing thou mayest loved be with equall crime.
     Spenser.

POPPIES & WHEAT

AUGUST.

Golden wheat and poppies-red
   Grow together, side by side:
Fruitful corn, with drooping head,
   Brilliant poppies open wide.
Not the brightest, not the proudest
   Are of greatest service here,
But the bowing heads of harvest,
   That with plenty crown the year.

AUGUST.

See now at work the reaper bands,
With lightsome heart and eager hands,
And mirth and music cheer the toil ;
While sheaves that stud the russet soil,
And sickles gleaming in the sun,
Tell jocund Autumn is begun.

Pringle.

GOLDEN Autumn now is come. 'Tis glorious
August, time of harvest ! The year has now
reached its fullest maturity, like a man in prime
of life, with health, wealth, and vigour. The .corn is
fully ripe, and in the fields the busy reapers are hard at
work from early morn till late at eve. The days are
perceptibly shortening, but the sky is blue and clear ;
and though the sun at noon is strong, there is a light
breeze during the day ; the heat is not so oppressive as
it was last month, and the evenings are delightful.

August was called by the Romans *Sextilis,* or sixth
month of their year. It had originally twenty-the

days. Julius Cæsar, in reforming the Calendar, gave it thirty days ; but, after his death, Augustus Cæsar conferred on it his own name, and, taking a day from February gave August thenceforth thirty-one days. Our Anglo-Saxon forefathers called it *Arn-monat,* Barn Month, from the filling of barns, *arn* signifying harvest.

In addition to the wheat, oats, and barley, which are harvested this month, the hops are now gathered. In the English counties in which hops flourish, the hop-gardens present a picture which may well rival for landscape beauty the vineyards of Italy itself. Yet, with all this beauty, we feel that the year is declining —nearly all the lovely flowers which charmed us in Spring and Summer are gone. We notice many changes on every side. Though we may observe a sort of second spring in the young shoots from many of the trees and bushes, and may still see pretty flowers, such as the red poppy or ox-eye daisy in the fields, or woodbine in the hedges, we cannot but be reminded that the year is drawing to a close, and that ere long we shall have weary Winter back again.

PRIDE AND THE POPPIES.

"We little Red-caps are among the Corn,
Merrily dancing at early morn ;
We know that the Farmer hates to see
Our saucy red faces, but here are we !

"We pay no price for our summer coats,
Like those slavish creatures, Barley and Oats ;
We don't choose to be ground and eat
Like our heavy-head neighbour, Gaffer Wheat.

"Who dare thrash us, we should like to know ?
Grind us, and bag us, and use us so ?
Let meaner and shabbier things than we
So stupidly bend to utility !"

So said little Red-cap, and all the rout
Of the Poppy clan set up a mighty shout ;
Mighty for them, but, if you had heard,
You had thought it the cry of a tiny bird.

So the Poppy folk flaunted it over the field ;
In pride of grandeur they nodded and reeled,
And shook out their jackets, till nought was seen
But a wide, wide shimmer of scarlet and green.

The Blue-bottle sat on her downy stalk,
Quietly smiling at all their talk,
The Marigold still spread her rays to the sun,
And the purple Vetch climbed up to peep at the fun.

The homely Corn-cockle cared nothing—not she,
For the arrogance, bluster, and poor vanity
Of the proud Poppy tribe, but she flourished and grew,
Content with herself and her plain purple hue.

The sun went down, and rose bright on the morrow,
To some bringing joy, and to others e'en sorrow ;
But blithe was the rich, rosy Farmer that morn,
When he went with his reapers among the corn.

He trotted along, and he cracked his joke,
And chatted and laughed with the harvest folk ;
For the weather was settled, barometer high,
And heavy crops gladden'd his practised eye.

" We'll cut this barley to-day," quoth he,
As he tied his white pony under a tree,
" Next the upland wheat, and then the oats :"
How the Poppies shook in their scarlet coats !

Aye, shook with laughter, not fear, for they
Never dreamed they too should be swept away ;
And their laughter was spite, to think that all
Their useful neighbours were doomed to fall.

They swelled and bustled with such an air,
The corn-fields quite in commotion were,

And the farmer cried, glancing across the grain,
" How these rascally weeds have come up again ! "

" Ha, ha ! " laughed the Red-caps, " Ha, ha ! what a fuss
Must the poor weeds be in ! how they're envying us ! "
But their mirth was cut short by the sturdy strokes
They speedily met from the harvest folks.

And when low on earth each stem was laid,
And the round moon looked on the havoc made,
A Blue-bottle propped herself half erect,
And made a short speech—to this effect :—

" My dying kins-flowers and fainting friends,
The same dire fate alike attends
Those who in scarlet or blue are dressed ;
Then how silly the pride that so late possessed

" Our friends, the Red-caps ! how low they lie
Who were lately so pert, so vain, and high !
They sneered at us and our plain array :
Are we now a whit more humble than they ?

" They scorned our neighbours ; the goodly Corn
Was the butt of their merriment eve and morn ;
They lived on its land, from its bounty fed,
But a word of thanks they never have said.

" And which is the worthiest now, I pray ?
Have ye not learned enough to-day ?
Is not the Corn sheafed up with care ?
And are not the Poppies left dying there ?

"The Corn will be carried, and garnered up
To gladden man's heart both with loaf and cup ;
And some of the seed the land now yields
Will be brought again to its native fields.

"And grow, and ripen, and wave next year
As richly as this hath ripened here ;
And we poor weeds, though needed not,
Perchance may spring up on this very spot.

"But let us be thankful, and humble too ;
Not proud and vain of a gaudy hue ;
Ever remembering, though meanly drest,
That USEFULNESS is of all gifts the best."

L. A. Twamley.

HARVEST HOME.

COME, sons of Summer, by whose toil
We are the lords of wine and oil ;
By whose tough labours and rough hands
We rip up first, then reap our lands.
Crown'd with the ears of corn, now come,
And to the pipe sing harvest home !
Come forth, my lord, and see the cart
Drest up with all the country art :
See, here a maukin, there a sheet,
As spotless pure as it is sweet ;

The horses, mares, and frisking fillies,
Clad all in linen, white as lilies ;
The harvest swains and wenches, bound
For joy to see the hock-cart crown'd.
About the cart, hear how the rout
Of rural younglings raise the shout,
Pressing before, some coming after,
Those with a shout, and these with laughter.
Some bless the cart, some kiss the sheaves,
Some prank them up with oaken leaves ;
Some cross the fill-horse, some with great
Devotion stroke the home-borne wheat ;
While other rustics, less attent
To prayers than to merriment,
Run after, with their breeches rent !
Well, on, brave boys, to your lord's hearth,
Glitt'ring with fire, where, for your mirth,
Ye shall see first the large and chief
Foundation of your feast, fat beef ;
With upper stories, mutton, veal,
And bacon, which makes full the meal ;
With sev'ral dishes standing by,
As, here a custard, there a pie,
And here all-tempting frumentie.
And for to make the merry cheer,
If smirking wine be wanting here,
There's that which drowns all care, stout beer ;
Which freely drink to your lord's health,
Then to the plough, the commonwealth ;
Next to your flails, your fanes, your fatts,

Then to the maids with wheaten hats ;
To the rough sickle and the crooked scythe,
Drink, frolic, boy, till all be blythe !
Feed and grow fat, and, as ye eat,
Be mindful that the lab'ring neat,
As you, may have their full of meat ;
And know, besides, ye must revoke
The patient ox unto the yoke,
And all go back unto the plough
And harrow, though they're hang'd up now ;
And, you must know, your lord's word's true,—
Feed him ye must, whose food fills you ;
And that this pleasure is like rain,
Not sent ye for to drown your pain,
But for to make it spring again.

Herrick.

BLACKBERRY
SEPTEMBER

Tho' woodbines flaunt,
      and roses grow
O'er all the fragrant bowers,
Thou needest not be
      ashamed to show
Thy satin-threaded flowers;
For dull the eye,
      the heart as dull,
That cannot feel how fair,
Amid all beauty, beautiful
    Thy tender blossoms are.

SEPTEMBER.

Go to the silent Autumn woods !
  There has gone forth a spirit stern ;
Its wing has waved in triumph here,
The Spring's green, tender leaf is sere,
  And withering hangs the summer fern.

Mary Howitt.

HARVEST Home ! What joy these words express ! The harvest, commenced last month, is nearly completed. What a picturesque scene is the harvest field : the yellow corn, the reaper with his sickle, the binders tying the golden sheaves, the gleaners following ; old women, young maidens, and little children, in dress of every tint of colour, form a picture that delights the heart of the artist. And now 'tis evening. The last load of corn is on its way to the stack-yard, followed by a motley group of merry rustics. They are about to enjoy their harvest-home supper, and a merry dance by the brilliant light of the glorious harvest moon. At no season does Cynthia shine so brightly as in harvest. Kirke White justly says :—

> " Moon of harvest ! herald mild
> Of Plenty, rustic labour's child,
> Hail ! oh, hail ! I greet thy beam,
> As soft it trembles o'er the stream,
> And gilds the straw-thatched hamlet wide,
> Where innocence and peace reside ! "

When the year began in March, September was the seventh month of the year—hence its name, September. It is now inappropriate, as the year commences two months earlier. Our Saxon ancestors called it *Gerst-monat* or Barley Month, from *gerst*, barley.

The decline of the year has now commenced. The leaves of the trees are donning their golden and tawny tints. The orchard trees are laden with pears, plums, and apples. The hedgerows are brightened with the scarlet berries of hips, haws, and honeysuckles ; as well as with the bright fruit of the privet, the thorn, the elder, and the blackberry. The harvest is over, and we cannot but feel thankful to the Giver of all good things for it, and for the many bounties which we now so freely enjoy.

THE BRAMBLE-FLOWER.

THY fruit full well the schoolboy knows,
　　Wild bramble of the brake !
So put thou forth thy small white rose,
　　I love it for his sake.
Though woodbines flaunt and roses glow
　　O'er all the fragrant bowers,
Thou needest not be ashamed to show
　　Thy satin-threaded flowers ;
For dull the eye, the heart as dull,
　　That cannot feel how fair,
Amid all beauty, beautiful
　　Thy tender blossoms are !
How delicate thy gauzy frill !
　　How rich thy branchy stem !
How soft thy voice when woods are still,
　　And thou sing'st hymns to them.
While silent showers are falling slow,
　　And, 'mid the general hush,
A sweet air lifts the little bough,
　　Love whispering through the bush !

The primrose to the grave is gone ;
   The hawthorn flower is dead ;
The violet by the mossed grey stone
   Hath laid her weary head ;
But thou, wild bramble ! back dost bring,
   In all their beauteous power,
The fresh, green days of life's fair Spring,
   And boyhood's blossomy hour.
Scorned bramble of the brake ! once more
   Thou bidd'st me be a boy,
To gad with thee the woodlands o'er,
   In freedom and in joy.

Ebenezer Elliott.

HARVEST HOME.

HERE once a-year Distinction lowers her crest ;
The master, servant, and the merry guest
Are equal, all ; and, round the happy ring,
The reaper's eyes exulting glances fling ;
And, warmed with gratitude, he quits his place,
With sunburnt hands and ale-enliven'd face,
Re-fills the jug, his honoured host to tend,
To serve at once the master and the friend ;
Proud thus to meet his smiles, to share his tale,
His nuts, his conversation, and his ale.

Bloomfield.

THE KNIGHT AND THE LADY FAIR.

"FORGET-ME-NOT."

TOGETHER they sate by a river's side,
  A knight and a lady gay,
And they watched the deep and eddying tide
  Round a flowery islet stray.

And, "Oh for that flower of brilliant hue,"
  Then said the lady fair,
"To grace my neck with the blossoms blue,
  And braid my nut-brown hair!"

The knight has plunged in the whirling wave,
  All for his lady's smile,
And he swims the stream with courage brave,
  And he gains yon flowery isle;

And his fingers have cropped the blossoms blue,
  And the prize they backward bear,
To deck his love with the brilliant hue,
  And braid her nut-brown hair.

But the way is long, and the current strong,
  And alas for that gallant knight!
For the waves prevail, and his stout arms fail,
  Though cheered by his lady's sight.

Then the blossoms blue to the bank he threw
  Ere he sank in the eddying tide!
And, "Lady, I'm gone,—thine own true knight,—
  Forget me not!" he cried.

This farewell pledge the lady caught,
 And hence, as legends say,
The flower is a sign to awaken thought
 For friends who are far away ;

For the lady fair of her knight so true
 Still remembered the hapless lot,
And she cherished the flower of brilliant hue,
And she braided her hair with the blossoms blue,
 And then called it " Forget-me-not ! "

Bishop Mant.

MORNING IN AUTUMN.

It was a fair and mild Autumnal sky,
And earth's ripe treasures met the admiring eye,
As a rich beauty, when her bloom is lost,
Appears with more magnificence and cost.
. . . .
Cold grew the foggy morn, the day was brief,
Loose on the cherry hung the crimson leaf ;
The dew dwelt ever on the herb ; the woods
Roared with strong blasts, with mighty showers the floods :
All green was vanished, save of pine and yew,
That still displayed their melancholy hue ;
Save the green holly, with its berries red,
And the green moss that o'er the gravel spread.

Crabbe.

WILD FLOWERS.

BEAUTIFUL flowers of the woods and fields,
  That bloom by mountain streamlets, 'mid the heather,
  Or into clusters 'neath the hazels gather,
Or, where by hoary rocks you make your bields,
  And sweetly flourish on through summer weather,—
    I love ye all !

Beautiful flowers ! to me ye fresher seem
  From the Almighty Hand, that fashioned all,
  Than those that flourish by a garden wall ;
And I can image you as in a dream,
    Fair, modest maidens, nursed in hamlets small—
      I love ye all !

Beautiful gems, that on the brow of earth
  Are fixed as in a queenly diadem !
  Though lowly ye, and most without a name,
Young hearts rejoice to see your buds come forth,
  As light erewhile into the world came—
    I love ye all !

Beautiful things ye are, where'er ye grow !—
  The wild red rose, the speedwell's peeping eyes,
  Our own bluebell, the daisy, that doth rise
Wherever sunbeams fall or winds do blow,
  And thousands more of blessèd forms and dyes—
    I love ye all !

Beautiful nurslings of the early dew !
   Fanned in your loveliness by every breeze,
   And shaded o'er by green and arching trees,
I often wish that I were one of you,
   Dwelling afar upon the grassy leas—
      I love ye all !

Beautiful watchers, day and night ye wake !
   The evening star grows dim and fades away,
   And morning comes and goes, and then the day
Within the arms of night its rest doth take,
   But ye are watchful wheresoe'er we stray—
      I love ye all !

Beautiful objects of the wild bee's love,
   The wild bird joys your opening bloom to see,
   And in your native woods and wilds to be !
All hearts to Nature true ye strangely move,
   Ye are so passing fair—so passing free—
      I love ye all !

Beautiful children of the glen and dell,
   The dingle deep, the moorland stretching wide,
   And of the mossy fountain's sedgy side,
Ye o'er my heart have thrown a lovesome spell ;
   And though the worldling, scorning, may deride—
      I love ye all !

                         R. Nicol.

WILD ROSE

OCTOBER.

When the chill October wind
Oft is sadly sighing,
When the leaves in circlets fly,
All seems dead or dying.
Wild Rose, then thy ruddy leaves,
And scarlet berries shining.
Give bright glow to bare hedgerow
And the year declining.

OCTOBER.

Autumn's sighing,
Moaning, dying,
Clouds are flying
   On like steeds ;
While their shadows
O'er the meadows
Walk like widows
   Decked in weeds.

Red leaves trailing
Fall unfailing—
Dropping, sailing
   From the wood,
That, unpliant,
Stands defiant,
Like a giant
   Dropping blood.

T. B. Read.

IS golden October. "The harvest is past, the Summer is ended." The lovely flowers of Spring and Summer are gone. The fruit trees have yielded up their fruit ; but the forest trees are now in their riches of golden glory, and the foliage is indeed lovely. The noble oak, the horse-chestnut, the elm, the beech, the ash, the lime, and the poplar, vie with each other in the brilliancy and beauty of their autumnal tints. The graceful firs and hardy evergreens still retain the summer shades of green. 'Tis evening. A sudden breeze springs up. It gradually increases in intensity towards night-fall, when it blows a full gale. The wind whistles among the trees, and the boughs are rudely shaken. The fallen leaves are caught up and whirled in eddying circles on the dry ground. At last the wind abates, and then comes a heavy fall of rain. The trees, in one night, are robbed of nearly all their golden tints, and thus announce the speedy return of Winter.

October was the eighth month of the ancient Roman Calendar—hence its name. Our Saxon ancestors styled it *Wyn-monat*, or *Wein-monat, i.e.*, Wine Month. The ancient Germans called it *Winter-fylleth*, from the approach of Winter with the full moon of the month.

Although, owing to the decay of nature, we cannot but have a sort of melancholy feeling about this month, nevertheless, we have occasionally in it some of the finest and most bracing weather of the year. There is often frost in the morning and evening, and warm sunshine in mid-day, accompanied by an exhilarating breeze ; and it is considered the very best time of the year to enjoy "a sniff of the briny" at the sea-side.

AUTUMN.—A DIRGE.

THE warm sun is failing,
The bleak wind is wailing,
The bare boughs are sighing,
The pale flowers are dying,
    And the Year
On the earth, her death-bed,
In shroud of leaves, dead
    Is lying.
Come, Months, come away,
From November to May ;
In your saddest array
Follow the bier
Of the dead, cold Year,
And, like dim shadows, watch by her sepulchre.

The chill rain is falling,
The night-worm is crawling,
The rivers are swelling,
The thunder is knelling
    For the Year ;
The blithe swallows are flown,
And the lizards, each gone
    To his dwelling.

Come, Months, come away,
Put on white, black, and grey ;
Let your light sisters play,
Ye follow the bier
Of the dead, cold Year,
And make her grave green with tear on tear.

Shelley.

AUTUMNAL SONNET.

Now Autumn's fire burns slowly along the woods,
And, day by day, the dead leaves fall and melt,
And, night by night, the monitory blast
Wails in the key-hole, telling how it pass'd
O'er empty fields, or upland solitudes,
Or grim wide wave ; and now the power is felt
Of melancholy, tenderer in its moods
Than any joy indulgent Summer dealt.
Dear friends, together in the glimmering eve,
Pensive and glad, with tones that recognise
The soft, invisible dew on each one's eyes,
It may be somewhat thus we shall have leave
To walk with memory, when distant lies
Poor Earth, where we were wont to live and grieve.

W. Allingham.

AUTUMN.

THE Autumn is old ;
    The sere leaves are flying ;
He hath gathered up gold,
    And now he is dying :
    Old age, begin sighing !

The vintage is ripe,
    The harvest is heaping ;
But some that have sowed
    Have no riches for reaping :
    Poor wretch, fall a-weeping !

The year's in the wane,
    There is nothing adorning,
The night has no eve,
    And the day has no morning ;
    Cold Winter gives warning.

The rivers run chill,
    The red sun is sinking,
And I am grown old,
    And life is fast shrinking :
    Here's enow for sad thinking !

Hood.

TO MEADOWS.

YE have been fresh and green,
 Ye have been filled with flowers,
And ye the walks have been
 Where maids have spent their hours.

Ye have beheld where they
 With wicker arks did come
To kiss, and bear away
 The richer cowslips home.

You've heard them sweetly sing,
 And seen them in a round ;
Each virgin, like the Spring,
 With honeysuckles crowned ;

But now we see none here
 Whose silvery feet did tread,
And, with dishevelled hair,
 Adorned this smoother mead.

Like unthrifts, having spent
 Your stock, and needy grown,
You're left here to lament
 Your poor estates alone.

Herrick.

AUTUMN FLOWERS.

THOSE few pale Autumn flowers,
  How beautiful they are !
Than all that went before,
Than all the summer store,
  How lovelier far !

And why?   They are the last !
  The last ! the last ! the last !
Oh ! by that little word
How many thoughts are stirred
  That whisper of the past !

Pale flowers ! pale, perishing flowers !
  Ye're types of precious things ;
Types of those better moments,
That flit, like life's enjoyments,
  On rapid, rapid wings !

Last hours with parting dear ones
  (That time the fastest spends),
Last tears in silence shed,
Last words half-utterèd,
  Last looks of dying friends.

Who but would fain compress
  A life into a day ?—
The last day spent with one
Who, ere the morrow's sun,
  Must leave us, and for aye ?

O precious, precious moments !
Pale flowers, ye're types of those !
The saddest, sweetest, dearest,
Because, like those, the nearest
To an eternal close.

Pale flowers ! pale, perishing flowers !
I woo your gentle breath ;
I leave the summer rose
For younger, blither brows ;—
Tell me of change and death !

Caroline Southey.

WITHERING—WITHERING.

WITHERING—withering—all are withering—
All of Hope's flowers that youth hath nursed,—
Flowers of Love too early blossoming,
Buds of Ambition too frail to burst !

Faintly—faintly—O ! how faintly
I feel life's pulses ebb and flow ;
Yet, Sorrow, I know thou dealest daintily
With one who should not wish to live moe.

Nay ! why, young heart, thus timidly shrinking ?
Why doth thy upward wing thus tire ?
Why are thy pinions so droopingly sinking,
When they should only waft thee higher ?

Upward—upward let them be waving,
Lifting the soul toward her place of birth ;
There are guerdons there more worthy thy having,
Far more than any these lures of the earth.—*Hoffman.*

CHRYSANTHEMUM

NOVEMBER.

When grey Winter comes apace,
And the Summer flowers are gone,
Golden brightness in thy face
Then we gladly look upon.
May we all thy teachings heed—
That, in paths of sombre care,
There is hope in darkest need,
And some brightness even there:
Golden rays of faith and worth,
Springing from the dust of earth.

NOVEMBER.

The drooping year is in the wane,
  No longer floats the thistle-down ;
The crimson heath is wan and sere,
The sedge hangs withering by the mere,
  And the broad fern is rent and brown.

Mary Howitt.

NOVEMBER,—pioneer of Winter, month of fog
and rain, of dirty days and dark nights,—is one
of the most unwholesome and uncomfortable
months of the year. The sun rarely shows his bright
face. The rivers are full to overflowing, and the
hedges and trees are nearly all quite bare. Yet, in the
country, the observer of nature may notice many
things which throw beauty and brightness over the
scene. If the richly-coloured foliage of the trees is
gone, we have many bright berries in the hedges,
which are now in full perfection : the blackberry, the
haw, the hip, the sloe, the cranberry, and the dark

fruit of the privet and the ivy. On the ditch sides we may see many graceful ferns, which make a pleasant show of bright green.

November was the ninth month of the Roman year —hence its name. The Saxons styled it *Wint-monat*, or Wind Month, from the gales of wind prevalent at this season.

If November weather be disagreeable in the country, it is still more so in town, particularly in London, where the thick, yellow fog is sometimes so dense as to turn noon into night, and impede all business and traffic through the streets. This cannot be better described than in the words of the famous wit and poet, Thomas Hood :—

No sun—no moon—no morn—no noon—
　No dawn—no dusk—no proper time of day—
　　No sky—no earthly view—
　　No distance looking blue—
　No road—no street—no "t'other side the way"—
　　No end to any row—
　No indications where the crescents go—
　　No top to any steeple—
　No recognitions of familiar people—
No warmth—no cheerfulness—no healthful ease—
　No comfortable feel in any member—
No shade—no shine—no butterflies—no bees—
　No fruits—no flowers—no leaves—no birds—
　　November.

NOVEMBER.

THE mellow year is hasting to its close,
  The little birds have almost sung their last ;
  Their small notes twitter in the dreary blast—
That shrill-piped harbinger of early snows.
The patient beauty of the scentless rose,
  Oft with the morn's hoar crystal quaintly glassed,
  Hangs, a pale mourner for the Summer past,
And makes a little Summer where it grows.
In the chill sunbeam of the faint, brief day,
  The dusky waters shudder as they shine ;
The russet leaves obstruct the straggling way
  Of oozy brooks, which no deep banks define ;
And the gaunt woods, in ragged, scant array,
  Wrap their old limbs with sombre ivy twine.

Hartley Coleridge.

THE WINDY NIGHT.

ALOW and aloof
Over the roof
How the midnight tempests howl !
With a dreary voice, like the dismal tune
Of wolves that bay at the desert moon ;
Or whistle and shriek
Through limbs that creak.
"Tu-who ! Tu-whit !"
They cry and flit,
"Tu-who ! Tu-whit !" like the solemn owl !

Alow and aloof
Over the roof
Sweep the moaning winds amain,
And wildly dash
The elm and ash
Clattering on the window-sash,
With a clatter and patter,
Like hail and rain,
That well-nigh shatter
The dusky pane !

Alow and aloof
Over the roof
How the tempests swell and roar !
Though no foot is astir,
Though the cat and the cur

Lie dozing along the kitchen floor,
  There are feet of air
  On every stair,
  Through every hall !
  Through each gusty door
  There's a jostle and bustle,
  With a silken rustle,
Like the meeting of guests at a festival.

  Alow and aloof
  Over the roof
How the stormy tempests swell !
  And make the vane
  On the spire complain :
They heave at the steeple with might and main,
  And burst and sweep
  Into the belfry on the bell !
They smite it so hard, and they smite it so well,
 That the sexton tosses his arms in sleep,
And dreams he is ringing a funeral knell !

T. B. Read.

WINTER SONG.

  SUMMER joys are o'er ;
  Flowerets bloom no more ;
  Wintry winds are sweeping ;
  Through the snow-drifts, peeping,
   Cheerful evergreen
   Rarely now is seen.

Now no plumèd throng
Charms the wood with song ;
Ice-bound trees are glittering ;
Merry snow-birds, twittering,
Fondly strive to cheer
Scenes so cold and drear.

Winter, still I see
Many charms in thee !
Love thy chilly greeting,
Snow-storms fiercely beating,
And the dear delights
Of the long, long nights.

Ludwig Holty.

WINTER.

WHEN icicles hang by the wall,
And Dick, the shepherd, blows his nail,
And Tom bears logs into the hall,
And milk comes frozen home in pail ;
When blood is nipp'd, and ways be foul,
Then nightly sings the staring owl,
"Tu-whit !
"Tu-who !" a merry note,
While greasy Joan doth keel the pot.

When all aloud the wind doth blow,
And coughing drowns the parson's saw,
And birds sit brooding in the snow,
And Marian's nose looks red and raw ;

When roasted crabs hiss in the bowl,
Then nightly sings the staring owl,
    "Tu-whit!
"Tu-who!" a merry note,
While greasy Joan doth keel the pot.

*   *   *   *   *   *

BLOW, blow, thou winter wind,
Thou art not so unkind
    As man's ingratitude ;
Thy tooth is not so keen,
Because thou art not seen,
    Although thy breath be rude.
Heigh-ho ! sing heigh-ho ! unto the green holly ;
Most friendship is feigning, most loving mere folly :
    Then, heigh-ho, the holly !
    This life is most jolly.

Freeze, freeze, thou bitter sky,
That dost not bite so nigh
    As benefits forgot :
Though thou the waters warp,
Thy sting is not so sharp
    As friend remember'd not.
Heigh-ho ! sing heigh-ho ! unto the green holly ;
Most friendship is feigning, most loving mere folly :
    Then, heigh-ho, the holly !
    This life is most jolly.

Shakespeare.

THE DEATH OF THE FLOWERS.

How happily, how happily the flowers die away !
Oh ! could we but return to earth as easily as they !
Just live a life of sunshine, of innocence and bloom,
Then drop without decrepitude or pain into the tomb !

The gay and glorious creatures ! they neither " toil nor spin :"
Yet, lo ! what goodly raiment they're all apparelled in !
No tears are on their beauty, but dewy gems, more bright
Than ever brow of Eastern queen endiademed with light.

The young rejoicing creatures ! their pleasures never pall,
Nor lose in sweet contentment, because so free to all !—
The dew, the showers, the sunshine, the balmy, blessed air,
Spend nothing of their freshness, though all may freely share.

The happy, careless creatures ! of Time they take no heed,
Nor weary of his creeping, nor tremble at his speed ;
Nor sigh with sick impatience, and wish the light away ;
Nor, when 'tis gone, cry dolefully, " Would, God, that it were
    day ! "

And, when their lives are over, they drop away to rest,
Unconscious of the penal doom on holy Nature's breast.
No pain have they in dying, no shrinking from decay—
Oh ! could we but return to earth as easily as they !

C. Bowles.

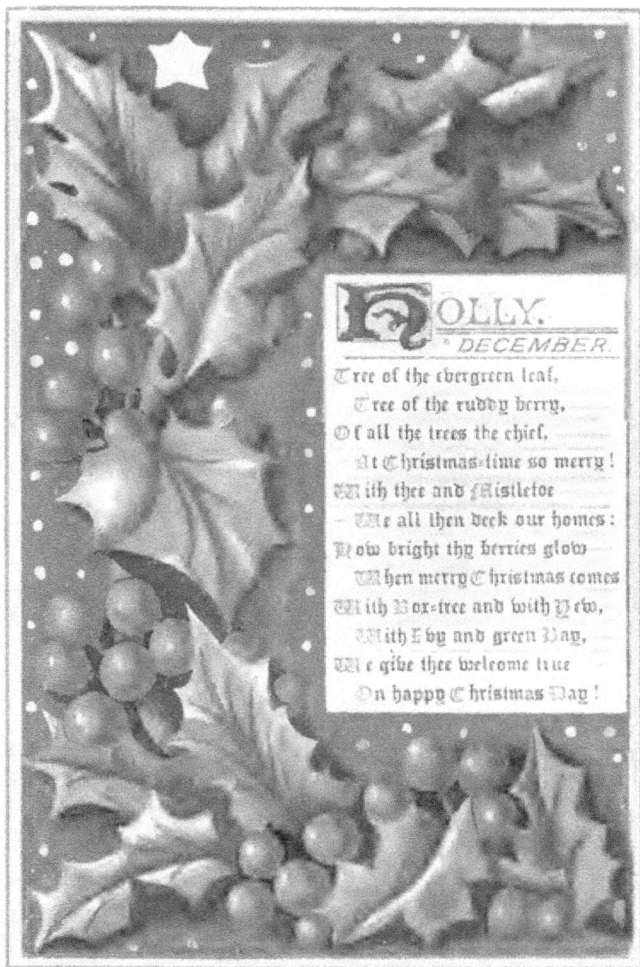

HOLLY.

DECEMBER.

Tree of the evergreen leaf,
  Tree of the ruddy berry,
Of all the trees the chief,
  At Christmas-time so merry!
With thee and Mistletoe
  We all then deck our homes:
Now bright thy berries glow
  When merry Christmas comes
With Box-tree and with Yew,
  With Ivy and green Bay,
We give thee welcome true
  On happy Christmas Day!

DECEMBER.

O Winter ! ruler of the inverted year,
Thy scattered hair with sleet like ashes filled,
Thy breath congealed upon thy lips, thy cheeks
Fringed with a beard, made white with other snows
Than those of age, thy forehead wrapped in clouds,
A leafless branch thy sceptre, and thy throne
A sliding car, indebted to no wheels,
But urged by storms along its slippery way.

From " The Task."—Cowper.

DARK December is here now—month of the shortest day and the longest night, of rain and wind,
of snow and ice. It is the month of our great
Christian festival, happy Christmas, the time of family
re-unions, of joyous greetings, and of welcome presents.
December is sometimes wet and windy ; but more
often, particularly in the latter part of the month, it is
frosty and snowy. What is more delightful than a
smart walk on a frosty December day ? The spotless

snow is on the ground, a thick sheet of ice is on the
ponds and lakes, and the trees are laden with snowy
crystals.   In-doors the enjoyment is even greater.
What a scene at Christmas-tide,

"Which now is returned, when we shall have, in brief,
 Plum pudding, goose, turkey, minced pies, and roast beef;"

with the Christmas games, the kisses under the mistletoe, and the merry dance, finishing the evening with
the ever-welcome "Sir Roger de Coverley."

December, like the preceding three months, derives
its name from the place which it held in the ancient
Roman Calendar—the tenth month in the year.   The
Saxons termed it *Winter-monat*, or Winter Month ;
but, after their conversion to Christianity, they named
it *Heligh-monat*, or Holy Month, in allusion to the
birth of our Saviour.

There is now little vegetation going on, except in
conservatories, under glass.   We have, nevertheless,
the "Christmas-rose," or black hellebore (a beautiful
flower), the yellow jasmine, and others, in full perfection.   The winter-flowering larustinus is in bloom, and
we have the holly, with its fruit of coral, and the sacred
mistletoe, with its berries of pearl, which glow so
brightly in the Christmas decorations in our homes
and our churches.

THE HOLLY TREE.

O READER ! hast thou ever stood to see
        The holly tree?
The eye that contemplates it well perceives
        Its glossy leaves,
Ordered by an Intelligence so wise
As might confound the atheist's sophistries.

Below a circling fence its leaves are seen,
        Wrinkled and keen ;
No grazing cattle through their prickly round
        Can reach to wound ;
But, as they grow where nothing is to fear,
Smooth and unarmed the pointless leaves appear.

I love to view these things with curious eyes,
        And moralize ;
And in this wisdom of the holly tree
        Can emblems see
Wherewith, perchance, to make a pleasant rhyme ;
One which may profit in the after-time.

Thus, though abroad, perchance, I might appear
          Harsh and austere ;
To those who on my leisure would intrude,
          Reserved and rude ;
Gentle at home amid my friends I'd be,
Like the high leaves upon the holly tree.

And should my youth, as youth is apt, I know,
          Some harshness show,
All vain asperities I, day by day,
          Would wear away,
Till the smooth temper of my age should be
Like the high leaves upon the holly tree.

And as, when all the summer trees are seen
          So bright and green,
The holly leaves their fadeless hues display,
          Less bright than they ;
But when the bare and wintry woods we see,
What then so cheerful as the holly tree ?

So, serious should my youth appear among
          The thoughtless throng ;
So would I seem, among the young and gay,
          More grave than they,
That in my age as cheerful I might be
As the green winter of the holly tree.

                              Robert Southey.

THE SNOW STORM.

ANNOUNCED by all the trumpets of the sky,
Arrives the snow, and, driving o'er the fields,
Seems nowhere to alight. The whited air
Hides hills and woods, the river, and the heaven,
And veils the farm-house at the garden's end.
The sled and traveller stopped, the courier's feet
Delayed ; all friends shut out, the house-mates sit
Around the radiant fireplace, enclosed
In a tumultuous privacy of storm.

Come, see the North-wind's masonry !
Out of an unseen quarry, evermore
Furnished with tile, the fierce artificer
Covers his white bastions with projected roof
Round every windward stake, or tree, or door ;
Speeding, the myriad-handed, his wild work
So fanciful, so savage ; nought cares he
For number or proportion. Mockingly,
On coop or kennel he hangs Parian wreaths ;
A swan-like form invests the hidden thorn.
And when his hours are numbered, and the world
Is all his own, retiring as he were not,
Leaves, when the sun appears, astonished art
To mimic in slow structures, stone by stone,
Built in an age, the mad wind's night-work—
The frolic architecture of the snow.

Emerson.

CHRISTMAS.

MERRILY, cheerily, ring out the chimes !
Christmas-tide is the most blessed of times !
Once more returned to us, Christmas is come,
Bringing sweet joy and peace to every home.

Christmas reminds us all of the glad morn,
When, as had been foretold, Jesus was born ;
Angels from hosts above herald His birth,
" Glory to God on high, peace be on earth."

He, from the vale of death, man came to save,
And gain the victory over the grave !
Great was the victory, great is our joy,—
Jesus, the Conqueror, sin shall destroy.

In that great victory let us rejoice,
Gladly sing praise to God with heart and voice ;
And for His gifts to us, mercy, and love,
Join in the praises of angels above !

Let there be everywhere innocent mirth,
Hang up bright evergreens over each hearth !
Ring out a merry peal, ring it out clear,
King of the Winter days, Christmas, is here !

                                   The Editor.

THE CHRISTMAS HOLLY.

THE rose it is the love of June,
  The violet that of Spring,
But all those faithless fading flowers,
  That take the South-wind's wing,
As craven blooms I hold in scorn,
The holly's the wreath for a Christmas morn !

Its berries are red as a maiden's lip,
  Its leaves are of changeless green,
And anything changeless now, I wiss,
  Is somewhat rare to be seen !—
The holly which fall and frost has borne,
The holly's the wreath for a Christmas morn !

Its edges are set in keen array :
  They are fairy weapons, bared ;
And, in an unlucky world like ours,
  'Tis well to be prepared.
Like helm on crest of warrior borne,
The holly's the wreath for a Christmas morn !

The holly it is no green-house plant,
  But grows in the common air ;
In the peasant's lattice, the castle hall,
  Its green leaves alike are there.
Its lesson should in mind be borne—
The holly's the wreath for a Christmas morn !

Anon.

NEW YEAR'S EVE.

" Ring out the old, ring in the new."

WHAT joyful sounds at noon of night
   Burst out upon the ear !
What sudden chimes, eight notes as one,
   Roll out both far and near !
On wings of wind those gladsome sounds
   Spread over land and sea—
The new year's come, the old year's gone,
   Lost in Eternity.

Ring out, wild chimes, ring out again !
   Thy midnight notes awake
Sad memories of joy and grief,
   Though dear for old year's sake ;
Of happy days for ever gone,
   And lost friends, ever dear,
Who even in their Heavenly home
   May greet the coming year.

Ring out again with might and main !
   Ring out another peal
To welcome in the glad new year
   With joy we can't conceal !
Oh ! may it bring us every good,
   And banish every ill ;
And when at last this new year dies,
   May we be happy still !

The Editor.

www.ingramcontent.com/pod-product-compliance
Lightning Source LLC
Chambersburg PA
CBHW021935190326
41519CB00009B/1025